GIANTS OF THE SEA

GIANTS OF THE SEA

GALLERY BOOKS
An Imprint of W. H. Smith Publishers Inc.

112 Madison Avenue New York NY 10016

THE IMAGE BANK®

111 Fifth Avenue New York NY 10003

First published in 1988 in New York by Gallery
Books, an imprint of W.H. Smith Publishers Inc.,
112 Madison Avenue, New York, N.Y. 10016

ISBN 0-8317-3905-3

For rights information about the photographs in
this book please contact:

The Image Bank
111 Fifth Avenue, New York, N.Y. 10003

Manufactured in Singapore

Produced by Robert M. Tod
Art Direction and Design by Mark Weinberg
Assistant Art Direction by Dana Shimizu Lee
Written by Ken Martin
Edited by Sheila Buff
Photo Research: Esperanza Arenas & Tamara Kahn
Editorial Assistance: Elizabeth Loonan

Contents

Almost three-quarters of the earth's surface is covered by the sea. Throughout its realm, sea creatures from microscopic to massive swim, drift and creep along, going about their daily business of staying alive.

But just how huge are these creatures that inhabit the sea? Accounts have accumulated through the ages of so-called giants of the sea. They range from frightening monsters like the Kraken of Sweden (now believed to have been a giant squid) that supposedly attacked great ocean vessels, to the friendly dolphins of ancient Greek legends, which allowed children to ride the waves on their backs.

Stories about giant clams, giant octopuses and more have become a part of the human race's family album, to be passed on from one generation to another.

But where does the truth lie? In reality, some 300 million years ago, the ancient seas did house a group of giant fish called *placoderms*. The largest placoderm ever found in the fossil record is the dinichthys, which had a mammoth skull and reached a length of 30 feet.

Ancient seas were also the residence of giant reptiles that gradually left the land to re-enter the ocean. One of these, the ichthyosaur, had a streamlined, fish-like body, but like all reptiles, it had to leave the water at intervals in order to breathe air. From fossil remains, scientists estimate that these creatures of 200 million years ago were nearly 20 feet long.

Another ancient sea monster descended from the early reptiles was the plesiosaur. Fossil bones tell us that the plesiosaur was 50 feet long, with a giraffe-like neck half its length.

Today, thanks to intensified scientific investigation, we find that many frightening sea creatures were no more than figments of overly fertile imaginations.

But not all.

The fact remains that there were—and still are—both peaceful and deadly giants living in the sea.

Blue Whales: Biggest of the Big

The giant of giants anywhere—on the sea or on the land—is the blue whale. In fact, the blue whale is so large that it needs the buoyancy of the water in order to exist. Put such a gargantua on land, and it would collapse under its own weight.

Blue whales, like all whales, are cetaceans. Like fish, cetaceans swim in the seas. But since these creatures are warm-blooded mammals, they need to breathe air. They also bear live young, and nurse and care for them. Whales share a common ancestry with sheep, cows, deer and giraffes. Yet, somewhere in the distant past, for reasons we can only guess at, these former land mammals returned to the sea, where in size the blue whale is second to none.

Blue whales have been found weighing 120 tons; that's even larger than the estimated weight of the biggest dinosaur that ever lived. Even a baby blue whale is a sight to behold, weighing almost three tons at birth.

Whales fall into two groups: baleen whales and toothed whales. Baleen whales have a substance called baleen or whalebone in their upper jaw, instead of teeth. This material acts like a strainer that keeps in the whale's favorite food—a shrimp-like creature called krill—while letting it spit out any water swallowed along with them. The blue whale is a baleen whale. Regardless of its enormous size, it subsists on one of the smallest animals in the sea.

Other huge baleen whales include the fin whale, the right whale, the sei whale, the Greenland right whale and that gentlest of giants, the humpback whale.

Humpback whales, which have been known to reach 40 tons, are most famous for their haunting "songs," which have endeared them to researchers. Recordings of the humpback's resonant sound patterns have even been sent into space along with human greetings on two of the four unmanned space vehicles bound for far beyond our solar system.

As for the toothed whales, the most famous is the whale immortalized by author Herman Melville as Moby Dick, the sperm whale. Although the sperm whale is smaller than the blue whale, it is in no way a lightweight. Ranked as the largest carnivore on earth, male sperm whales average 33 tons in weight and 47 feet in length. They are also the deepest divers of air-breathing animals; they can reportedly plunge up to 3,000 feet below sea level in their quest for food.

The sperm whale's favorite meal consists of giant squid. In fact, much of what we know about the giant squid comes from the remains of them recovered from the stomachs of dead sperm whales.

Other toothed whales include porpoises and/or dolphins. While some use the term interchangeably, others differentiate between the two large sea-faring mammals, based on the shape of their snouts. (A dolphin's snout is more pointed than a porpoise's snout.) The sleek, powerful killer whale is considered by many scientists to be the world's largest porpoise.

Killer whales, with their powerful jaws and up to 26 sharp, large teeth, were long feared as people eaters. This is no longer thought to be the case, although killer whales do use their teeth in the water on such prey as polar bears, seals, sea lions, walruses and whales of all sizes, even blue whales.

Even blue whales?

Yes, even blue whales. The killer whale's success as a hunter is easier to understand when you realize that these sea mammals travel in hunting packs, rather than tackle the giant of giants one-on-one.

Today, in oceanariums around the world, these world-class swimmers—known to reach estimated speeds of 30 miles per hour—prove themselves to be playful, intelligent show mammals. Sea World's famous Shamu is a killer whale. However, even in play you have to be careful with friends so huge. During a recent Sea World show, an animal trainer was seriously injured when a six-ton killer whale landed on him by accident.

Best known to humans are bottle-nosed dolphins; the ones with the wiseguy grins we see in aquatic shows. Dolphins are sociable creatures, known to come to the aid of other dolphins, as well as people in distress.

Like the large whales, dolphins were once killed in great numbers by humans. And also like whales, they are being protected now by new laws to stop the killing, and replenish Earth's supply of this fascinating creature who has much to teach humans, as well as to learn from them.

Some scientists are especially interested in trying to communicate with dolphins. Meanwhile, naval specialists are primarily concerned with the dolphin's echo-location skills; dolphins are able to identify objects miles away by the echoes of clicking sounds the dolphin makes.

The Largest Fish on Earth: The Whale Shark

While the ingenious, irresistible dolphins have been on earth for 25 million years, sharks have been in earth's waters for 250 million years. Before dinosaurs existed, there were sharks. And after dinosaurs became extinct, the sharks remained, as they do now.

Sharks differ from most other fish since fish normally have skeletons made of true bone, while a shark's skeleton is composed of a softer material, much like that forming the bridge of your nose. This material is called cartilage.

Today we know of approximately 350 species of sharks (only ten years ago, only 250 sharks were identified). The majority of them are predators. Only three species are plankton eaters: the whale shark, the basking shark, and the recently discovered "Megamouth," with apparently luminescent tissue in its mouth.

The whale shark, the largest known fish on earth, is similar to the whale in size as well as in feeding habits. Like the blue whale, this creature—capable of reaching 60 feet in length—has a strainer-like device at its gill, which lets it spit out any water it swallows along with its food.

Yet this giant of the sea, with a mouth containing 3,000 tiny teeth, is a docile inhabitant of open tropical waters. The whale shark is most dangerous to people when it accidentally hits a boat or whacks a diver with its powerful fins.

The basking shark is another plankton eater. This Sleeping Beauty of the Seas, the second biggest shark of them all, feeds on plankton until winter comes and food gets sparse. Then he drops to the ocean floor, and stretches out for a long snooze until spring comes, and a fresh supply of plankton awaits him.

Like some, but not all other sharks, the great white shark—which can reach lengths of 20 feet or more—is warm-blooded. This unique feature gives certain sharks (like bluefin tunas, another giant among fish, and also warm-blooded) more swimming power than ordinary fish. The great white can navigate great areas of the ocean in huge bursts of speed. It is also down on record as attacking more swimmers than any other shark.

In second place as a people-attacker is the tiger shark, so named for the tiger markings found on newborns of the species. The biggest tiger shark ever reported being captured was 1,780 pounds and found off the coast of South Carolina in 1964. Tiger sharks eat all kinds of sea creatures, including other sharks, as well as any sea birds they can catch.

Do sharks larger than the whale shark exist today?

Perhaps. We do know that 15 million years there existed a shark so monstrously huge that each of its teeth was as large as a human hand. The reconstructed jaw of that giant—called Carcharodon megalodon—is on exhibit at the American Museum of Natural History in New York City. However, not too long ago, teeth of a Carcharodon megalodon were discovered—only, unlike teeth found before, these weren't fossilized!

Could the Carcharodon megalodon be alive today?

More Giants Among Fish

The sea is full of mysteries—as well as other giant fish. Among fish with bony skeletons the sluggish ocean sunfish (mola) can reach as much as two tons in weight. Found primarily in the open waters of the Atlantic, Pacific, and Mediterranean seas, this leathery-skinned fish that appears to be all head is a rarity. It is also considered deadly for humans to eat. As for its own meals, the ocean sunfish is a master of the inhale technique; it simply inhales any small prey passing by.

Groupers, too, have the ability to inhale their meals. They create a suction that draws in their victims. In fact, famed ocean explorer Jacques Cousteau, encountering an enormous 13-foot-long grouper-type creature while aboard his famous ship *Calypso*, mused that perhaps this was the origin of the Jonah and the whale story, since no whale could swallow a person whole, but a giant grouper might.

Two more of the largest fish in the sea are the oarfish, which is often over 20 feet, and the ribbonfish (a kind of oarfish). These eel-like fish inhabit the deep ocean, but occasionally they may break through to the surface, looking all the world like mythological sea serpents come to life. In 1963 one was reported off Asbury Park, New Jersey, that appeared to be 50 feet long.

Among game fish, the largest on record is the black marlin. A giant of this species, weighing 2,250 pounds, was once harpooned off Cabo Blanco, Peru. An average good catch might weigh 400 to 500 pounds. Other varieties of marlin are also large enough to be considered prize game. The International Game Fish Association, which has strict rules for registering record-breaking catches, lists the largest Atlantic blue marlin at 1,282 pounds, and the largest Pacific marlin at 1,153 pounds. As for the striped marlin, this impressive jumper has been known to reach 417 pounds.

Another much sought-after trophy for sportspeople is the bluefin tuna. Found in most warm and temperate waters, these half-a-tonners have been nabbed by anglers with just a rod and reel. The largest bluefin tuna on record was caught off Nova Scotia in 1979 and weighed nearly three-quarters of a ton.

Swordfish and sailfish are two more very popular game fish. Both possess a powerful spear that requires they be handled with extreme care. While both fish look similar, they can be told apart with ease; the sailfish's dorsal fin (the one on its back) is more well developed and much brighter than that of the swordfish.

Swordfish on the average reach 11 feet, and 100 to 300 pounds. Yet they can grow much larger. World records list the largest swordfish as weighing 1,182 pounds. This 14-foot giant was caught off Chile in the 1950s.

The Atlantic sailfish is half the size of the Pacific sailfish, which averages 125 pounds. Both varieties enjoy swimming close to the surface, as swordfish do. Occasionally, they can be seen making spectacular leaps out of the water. Sailfish tend to be more peaceful than swordfish, which have been known to spear small boats, even when not provoked. But both sailfish and swordfish are an exciting challenge to sportsmen, since they put up a strong struggle when hooked.

Then, too, there's the giant sea bass. The record for this creature is 563 pounds and 72 inches on a catch found off Anacopa Island, California in 1968.

Manta Ray: The 3,500-Pound Devilfish

Rays, like skates, are related to sharks, since the skeletons of all three fish are comprised of the strong, elastic tissue called cartilage. When rays are resting, they look like sea pancakes, because they lie flat against the ocean bottom. When rays swim, they seem to "fly" through the water.

The largest ray is the manta ray, which can weigh a good ton and a half. Found in tropical and subtropical water, this bat-like sea giant has a huge, cavernous mouth located in front of its head, not underneath as in most rays.

However, the manta ray's bark is worse than its bite. In fact, the manta ray makes a sound like a bark, and when it skims through the water and belly flops for a landing, the noise of those 3,500 pounds breaking through the sea's surface can be heard for miles around. Yet manta rays enjoy a bland diet of tiny animals, and are mostly harmless to people.

Not stingrays, however. Over 30 kinds of stingrays exist. The most immense ones sport foot-long barbs on a jagged spine in the center of a whiplike tail. Perhaps the spine is the source for ancient Greek accounts of sea monsters flailing poisoned spears. Rays use their barbs to inject victims with a strong poison that can cause pain, and even death.

Stingrays weighing 400 pounds have reportedly been found in the Atlantic ocean. The eighteenth-century English navigator Captain Cook supposedly caught several huge Australian stingrays weighing up to 700 pounds during his Pacific travels.

Moray Eels: Rattlesnakes of the Sea

Eels are snake-like fish that live in fresh water as well as in the sea. Freshwater eels (which grow no bigger than eight feet) are known for their unusual migratory pattern: when it is time, every adult returns to a breeding ground near the Bermuda Islands, where they mate, lay their eggs, and die. Less known to us—and more dangerous—is the ocean's moray eel. While most moray eels are under five feet long, a number have been found to reach 10 feet or more.
And, for the record, about 50 years ago, European scientists discovered an extraordinarily

long deep-sea larva, or newly hatched eel. Unlike usual eel larvas, which measure two to four inches, this find was no less than six feet. That suggests that somewhere in the sea, giant adult eels of 60 feet exist!

As for moray eels, they live primarily in the warm, shallow waters around coral reefs. Scuba divers who swim among the reefs to catch glimpses of the colorful flamefish, bluehead, sergeant-major damselfish, imperial angelfish and more, are warned to be careful. Morays are known to hide in narrow crevices, and while usually shy, they will bite a hand or foot near a hole where they live. The result can be deadly, since the moray's strong grip can prevent a swimmer from getting free before he drowns. No wonder these mean-spirited critters are called the "rattlesnakes of the sea."

Giant Sea Turtles

A giant with a far better disposition than the moray eel is the giant sea turtle. Although turtles are identified by their hard shell, the largest of the five types of sea turtles is the only one without a shell. Instead, the leatherback is covered with a thick, brown leathery skin. This giant has been known to grow over seven feet, and weigh a ton or more.

The green turtle, which is only half the size of the leatherback, is still considered a giant of the sea.

The only vegetarian among the sea turtles, green turtles can reach 500 pounds in weight. Like the leatherback, the green turtle is a fast and graceful swimmer. The large front flippers on these creatures are actually fins, allowing them to clock speeds as fast as 20 miles an hour.

Monster Mollusks

Clams are part of a huge group of animals called mollusks, and while most of us think of clams as those little things we make chowder from, giant clams exist as well. The largest clam on earth is the Tridacna or giant bear's claw of the South Pacific. A Tridacna clam weighing 580 pounds sets the world record so far.

The giant clam gets excellent camouflage from the small plants and animals that grow on its shell. However, Pacific islanders hunt these colossal clams for their adductor muscle (the muscle that opens and closes the shell), which makes a delicious chowder, as well as for their shells, which are made into hunting implements.

Octopuses are mollusks without shells. While ancient legends have octopuses grasping great sea vessels in their writhing tentacles, the truth is that the octopus is a rather shy underwater chap. Octopuses seldom leave home, except to hunt or mate. Even after the male has performed his mating dance and impregnated his sweetheart, each octopus goes its separate way, returning home alone.

For many marine biologists, an octopus holds a special allure. It can blend in with its surroundings by changing not only its color, but even the texture and shape of its skin. And in the open sea, it can do its famous Houdini act, disappearing in a puff of the inky black fluid it discharges. Just as interesting to scientists is that this bulbous creature with the huge, haunting eyes has an intelligence superior to all other sea dwellers, save for the large mammals.

The largest octopus known is a species in the Pacific that reaches a weight of 125 pounds and a length of 20 feet, measuring across from one arm tip to the opposite arm tip.

However, larger octopuses are believed to exist. Recently, scientists were able to use new methods to measure the remains of a gigantic specimen washed ashore on a Florida beach in the 1890s. Their conclusion? The remains belonged to an octopus that must have been 200 feet across, and six tons in weight!

The octopus belongs to the same family as the cuttlefish, the chambered nautilus and the squid. In legend, the giant squid is described as a villainous tangle of tentacles that attacks sailors even up at the top of their ship's rigging. In reality, the giant squid—favorite food of the sperm whale—seldom comes to the surface of the sea. The largest of the known squid are 55 feet long from tip of tentacle to end of tail, and over four tons heavy.

The squid, with its 10 waving arms and the biggest eyes in the animal world, is the largest backboneless animal on earth. Strange, but true, this is the big brother or sister to the tiny squid that Italians cook up as calamari and Japanese serve as ika. The giant squid, although a thousand times the size of its tasty relatives, is not a thousand times as delicious. In fact, it doesn't taste very good at all.

A Variety of Other Sea Giants

Wherever you look, more giants of the sea are found. Surprisingly enough, the longest in length of all animals is not the blue whale, but the pink jellyfish. This creature's tentacles sometimes reach twice the length of the blue whale, or 200 feet.

Of the 300 kinds of jellyfish, about 200 have stingers that humans can feel. The man-of-war jellyfish has the largest stinger of his brethren, and a great many of them. Humans have been known to faint from the pain of its sting. But this fascinating creature is not one jellyfish. Rather, it's a whole colony of jellyfish all working together.

The parasol part of the man-of-war we see sailing along on the water is actually one kind of jellyfish. Beneath the parasol is a long digestive tube to which many individual jellyfish tentacles—or polyps—are attached. Each polyp, like each member of any community, has a special task to do. Some polyps catch food, others eat it, while still others are nursing polyps from which new jellyfish will grow.

The largest jellyfish of all is the cyanea capillata, or sea blubber. It may have as many as 800 tentacles, and lives primarily in the icy-cold waters of the Arctic and Southern seas. Smaller versions of this giant are found along both coasts of the United States. Whether large or small, the Cyanea has a dangerous venom.

The sea is also home to giants among starfish. The largest found thus far was just over 54 inches from arm to arm. This fascinating creature has the ability to sprout new limbs to replace any that are lost. Just as amazing is the starfish's ability to turn its stomach inside out and digest a tasty meal outside its body. Starfish are part of a group called echinoderms, which means spiky skins. They represent a mixture of both primitive and advanced evolution.

Then there are the crabs. Crabs are crustaceans, like the lobster, shrimp and many smaller creatures, some almost microscopic. In turn, crustaceans belong to a large family to which most of the world's animals also belong. These are the arthropods, distinguished by their jointed legs.

The giant among crustaceans is the Japanese spider crab. This crusty fellow reaches a length of

11 feet from the tip of one claw to the tip of the other, weighs as much as 40 pounds, and has spidery legs that reach heights of 12½ feet. It is found in the deep offshore waters of the North Pacific.

The crab's cousin, the lobster, has the interesting ability to shed its outer shell and grow another one each time its body weight increases by 15 percent. The largest lobster on record was nearly 45 pounds, found off Nova Scotia.

Manatees: Massive Mermaids

Move down the east coast of North America to Florida, and you'll find the sanctuary for one of the most peculiar water mammals ever, the manatee. Like whales and porpoises, manatees are mammals that never leave the sea, not even to bear their young. Manatees belong to one of the 19 mammal orders, the sirenian.

These unusual-looking creatures may well by the source of mermaid legends. When a sirenian surfaces to breathe, which it must do often, almost half its body comes straight up out of the water, as if it were standing. Even more suggestive, the female sirenian holds her baby to her breast, like a human mother does.

Poor Christopher Columbus. When he was sailing off the South American coast, he and his crew thought they saw three mermaids when he spotted these creatures. Much to his shock, they were nowhere as lovely as he'd expected.

Manatees were once heavily hunted for their tasty, veal-like meat, as well as for their skin and oil. One kind of sirenian, Stellar's sea cow, was hunted to extinction by whalers and sealers in its native Bering Sea.

Today, thanks to sanctuaries, such as the one in Crystal River, Florida, the once dwindling number of manatees is on the rise again. Scientists believe that manatees are distant cousins of the elephant who left the land for water millions of years ago.

The Seal: Another Great Mammal of the Sea

Like the manatee, those massive mammals, the seals descended from land mammals. Scientists know of 34 different species of seals, all of them fascinating. Some stay in one place from birth to death. Others migrate with the seasons, navigating great distances in huge herds. Unlike manatees, which are monogamous, adult male seals have harems. Also unlike manatees and whales, seals raise their young on land, even though they live mostly in the sea.

Seals range in size from the 85-pound ringed seal to the 8,500-pound southern male elephant seal. Like other seals, elephant seals were once hunted for their fine oil. But today, their number is increasing as new laws make hunters refrain from killing these giants, whose name comes from the nine-inch trunk of sorts attached to their snouts.

"Seal" is a term used for both seals and sea lions. True seals, like elephant seals, are distinguished by having no outside ear. Eared seals have visible ears. Sea lions are eared seals.

Those funny seals that steal the show at circuses are usually California sea lions, which are great performers. And the largest of the sea lions is Stellar's sea lion. Sea lions are able to traverse the land more easily than true seals. Also, a sea lion's coat is heavier than that of a true seal's.

Both true and eared seals belong to the mammal group known as pinnipeds. The third member of this group is the walrus. A walrus is a gigantic animal that can grow to 3,000 pounds. They are excellent divers, able to descend to great depths in search of their food, which is limited in size to what they can fit between their ivory tusks.

More Giants . . . For Sure!

A universe of discoveries still awaits us as we delve deeper and deeper into the treasure chest of the sea. The biggest fish used to be found on the surface of the water. But this is changing, as more sophisticated scuba gear and deep-diving submersibles increase our knowledge of the over 70 percent of the earth that is covered with water.

But then, records are meant to be broken. And as for giants? They have been with us for ages, and are with us still . . . to give us a little fright perhaps, as well as a great deal of wonder.

Sand-strewn caverns, cool and deep,
Where the winds are all asleep;
Where the spent lights quiver and gleam;
Where the salt weed sways in the stream;
Where the sea-beasts rang'd all round
Feed in the ooze of their pasture-ground;
Where the sea-snakes coil and twine,
Dry their mail and bask in the brine;
Where great whales come sailing by,
Sail and sail, with unshut eye,
Round the world for ever and aye.

—Matthew Arnold

Sperm whales are famous for their great diving expertise. These colossal creatures can stay under water for up to two hours. Hunted for their fine oil for centuries, the behavior of sperm whales remains a mystery to scientists.

Belugas, like narwhals, are smaller whales that normally stay near the Arctic. Adult belugas are completely white, while newborns are more pinkish-brown and gray.

Killer whales have long been feared as cold-blooded man-aters. Recent research contradicts this, although pods of killer whales do seek out such massive prey as seals and other whales. Killer whales are one of two toothed whales—the other is the sperm whale.

Until recently, data about humpbacks came from whaling records and whales washed ashore.

Now scientists use sophisticated research techniques to learn about living whales in their natural environment, the sea.

Early each winter, the California gray whale travels south from the Arctic Ocean to Baja California, where the female bears her young. During the journey south, they eat almost nothing, relying on their fat reserve from a summer fare of krill up north. The calves mature quickly; just two months after birth, they must begin their long odyssey back north.

Krill is the staple food for the humpback whale. Water along with krill will be taken in by the whale as he swims along. However, the baleen in his mouth—a kind of strainer—will separate the krill from the sea water, and spit the latter back out.

The gentlest of giants, humpback whales weighing 800 times the weight of a human being appear to be floating islands. Their hide houses an assortment of barnacles, algae and parasites.

A whale exhales off the coast of Gloucester, Massachusetts. In the old whaling days, a lookout was posted at the top of the mast; when he spotted a whale spouting, he would shout, "There he blows!"

Whales breathe through slits in the top of their heads called blowholes. The spray we see after a whale has come up for air is caused by his warm breath condensing in the cold air.

True giants of the sea, whales are found around the world. Some whales migrate in the winter, while others remain in one place all their lives. The basking whale, for instance, will hibernate at the bottom of the sea each winter until a more plentiful plankton supply is available come spring.

Dolphins are sociable animals with a strong sense of community. When one dolphin is hurt, or sick, other dolphins will carry him up to the surface of the water to help him breathe. They have also been known to rescue humans in the sea, and even to lead ships safely into harbor.

Dolphins are known to be intelligent. Not only do they seem to have a language of their own, but they can also imitate human speech.

Porpoise or dolphin? Many people use the two words interchangeably for these small whales. But purists may only label "porpoise" those animals with rounded snouts and teeth that are spade-shaped instead of conical.

The bottle-nosed dolphin keeps a tight hold on any slippery fish between its beaklike jaw and teeth. However, it does not use its teeth for chewing.

Since dolphins must take in a lot of oxygen, the duration of their dives is limited. A dolphin can hold his breath about as long as a human can.

With his mouth open and his triangular teeth showing, a shark makes it clear he's the boss. Every kind of shark has a unique type of teeth. Most have four or more rows of these pearly whites, with the teeth in the row behind pushing out the front teeth every two weeks.

Sharks traveled through the waters on Earth long before man existed. Perhaps they will do the same long after man is gone.

The whale shark is the biggest fish in the world. Although nicknamed "White Death" by some, whale sharks are not likely to attack a human. In fact, swimmers have been known to safely hitch rides on the back of these 50-foot, 15-ton gargantuans!

Strange-looking hammerhead sharks have long, hammer-shaped heads with an eye on each end. Large hammerheads are considered dangerous. Their favorite food is stingrays.

Tiger sharks are ferocious creatures that are known to reach lengths of 18 feet. These loners are identified by their tiger stripes, which are more noticeable in young sharks.

Sharks use all their senses to seek out food. Their sense of smell is especially sharp; they can catch a whiff of food miles away.

The bull shark is unique among sharks because it can swim from salt water to fresh water without harm. It is considered dangerous, no matter where it is found.

With jagged teeth about the size of your thumb, a great white shark is best seen in captivity, like this one at the New York Aquarium. It is known as a man-eater, since it attacks more people than any other shark on record, yet great whites usually don't find human flesh very tasty, and spit out their prey after one bite.

Slow, brownish nurse sharks of the tropical seas can reach 14 feet in length. Since they can breathe without moving, they are often found lying partially concealed on the bottom of the ocean.

This nurse shark has a friend: a remora fish. Remoras travel with sharks, using a suction disc on their heads to fasten themselves to their host. They dine in style on the scraps sharks leave over after eating, and in return they pick off tiny parasites from the shark's skin.

Sharks can see better when the light is dim. Perhaps that is why they search for food primarily when dawn breaks, or the sun sets, as it is here, off Mauritania in western Africa.

Giant groupers live along the sea bottom in the tropics. While there is no proof that a grouper ever swallowed a diver, pearl divers in the South Pacific claim this is possible. A large grouper can be 12 feet long, and weigh half a ton.

Groupers use their enormous mouths as suction pumps, creating a vacuum that sucks in creatures swimming by.

A black marlin swimming off Mauritius in the Indian Ocean. The average black marlin catch weighs 400 to 500 pounds, but one harpooned off Cabo Blanco, Peru weighed 2,250 pounds.

The master of the swift hit-and-run attack, the barracuda can strike without warning, then quickly disappear into the water. Known as "the tiger of the sea," this slender fish has powerful jaws and can reach six feet or more.

Rays, like skates, are related to the shark. All three are composed of cartilage, rather than bony material.

Stingrays range from hand-size to nearly 12 feet long and seven feet wide. They will not attack people unless provoked, but can be angered unintentionally, since they bury part or all of their bodies in mud or sand.

A butterfly ray with distinctive blue markings skims along the bottom of the Red Sea.

Two moray eels, the "rattlesnakes of the sea," peek out from under a starfish. Although most moray eels grow no longer than five feet, a number have been discovered that are over 10 feet. Morays can be retiring at times, aggressive at others.

Known in Latin as muraena helena, *the moray eel is found in coral crannies, where he can cause jagged wounds to unsuspecting divers who stick a hand or foot in his home.*

A network pattern decorates the skin of this resident of the reefs—a reticulated moray—who must keep his mouth open at all times in order to breathe.

The sea turtle's ancestors were gigantic prehistoric creatures. Today, in warm waters around the world, five species make up the sea turtle family. The largest are the leatherbacks, which can weigh a ton.

Hawksbill turtles are named for their upper jaw, which hooks down like a hawk's beak. Like all other sea turtles, except for the giant leatherback, the hawksbill has a back shell too thick for most animals to bite through. It is baby turtles that are most endangered, as they make their innocent trek from beach to sea.

Although sea turtles spend almost all their time in water, the female takes "shore leave" to nest and lay her eggs, which could pass for ping-pong balls. Sea turtles always return to nest on the same beach where they were born.

The sea turtle's strong flippers make him a world-class swimmer. He primarily uses his two large front flippers, while his two back flippers act as rudders and brakes.

The common octopus is extremely hard to keep in captivity. By elongating its body, the octopus can squeeze through narrow openings, much to the frustration of aquarium curators. The octopus is also good at other feats, such as changing color at will.

Each of the octopus' eight arms or tentacles is lined with suction cups. These cups help the octopus grab its prey and pull it to its parrot-like beak.

Screaming divers being carried off by a flailing eight-armed cousin to the squid are more fiction than fact. An octopus is really a shy creature who mostly hides among rocks. Its home has a barricade in front of it, and even after octopuses mate, they go their separate ways, to continue to live alone.

When in the water, the man-of-war looks like a floating balloon. But the balloon is actually one jellyfish, and under the water are the other parts; each tentacle attached to the balloon has an individual task vital to the whole organism.

While we don't think of starfish as being carnivorous, they are just that. A starfish will dine on dead and alive mollusks and crustaceans by turning its stomach inside out and dissolving its victim in the starfish's own digestive juices.

Looking like a magical dual lightbulb, this red-eye jellyfish drifts along in the water. The most dangerous of all jellyfish is the sea wasp found off the coast of Australia. The venom from this deadly creature can paralyze a person's heart before he makes it back to shore.

Starfish creep along the beach and sea floor to find their food. The hundreds of suction discs on their feet grip and release, keeping the starfish in motion.

Commonly found washed up on beaches, the heavily armored horseshoe crab can be frightening to see, yet it cannot bite or sting. Furthermore, it is not a true crab; it belongs to the same group as the now-extinct sea scorpion, which grew to 12 feet long.

The giant clam of the South Pacific, called the Tridacna, can grow to half a ton or more. The two-shelled mollusk with green algae growing inside its shell is not a man-eater, but it can trap an unwary diver's leg.

Human beings are fascinated by these 1,000-pound vegetarians. From November to March every year, divers, photographers, and tourists track down manatees in their winter haunts, like Crystal River. Unfortunately, the propeller blades of passing tourist boats have been known to leave permanent scars on the skins of these gargantuans.

Manatees are docile giants who were once killed in great numbers for their veal-like meat. Today they are protected in sanctuaries, like this one in Crystal River, Florida.

The remote Arctic habitat of Weddell seals has offered protection from hunters to these expert divers. (One Weddell seal made a controlled dive that lasted 45 minutes.)

Weddell seal pups nursing on their mother's rich milk can gain as much as seven pounds a day.

Tourists visit Ano Nuevo State Reserve in California in increasing numbers. Here elephant seals are free to live in peace.

Male elephant seals utter loud, gutteral sounds when any member of their harem of female seals and pups is threatened by the attentions of a challenging male. Usually no battle ensues; the loudest male wins.

Once killed in great numbers for their oil, laws to protect them are replenishing the world's supply of elephant seals. The seals' own lethargic nature made them prime targets for hunters.

The large, dark-brown California seals found along North America's Pacific coast are tremendous show-offs. They are the natural performers we see in circuses and aquarium shows.

The giant among seals is the elephant seal. The males can weigh an average of over two tons.

The harp seal's name comes from the harp-shaped black markings found on the adult's back. Hunters are most interested in the infants with their snowy white fur, although the pelts, blubber and meat of adults are also prized.

The crabeater seal has a misleading name, since it mainly eats the small shrimplike crustaceans called krill, not crabs.

While walruses walk on all fours, like their eared-seal brethren, they use their fishtail-like rear flippers to swim, like true seals. A male walrus can weigh a hefty 3,000 pounds.

Seals are known to be excellent divers. One northern elephant seal reportedly dove 2,900 feet.

INDEX OF PHOTOGRAPHERS